Study Guide for the
SCAT®

School and College Ability Test®
Elementary Series

SAMPLE QUESTIONS FOR GIFTED 2ND AND 3RD GRADERS

Table of Contents

Introduction*

The **School and College Ability Test (SCAT)**, is a standardized test conducted in the United States that measures math and verbal reasoning abilities in gifted and talented children. In most school districts children are screened using an above grade average test.

Students in grades 2-3 take the Elementary SCAT designed for students in grades 4-5. Students in grades 4-5 take the Intermediate SCAT designed for students in grades 6-8. Students in grades 6 and above take the Advanced SCAT designed for students in grades 9-12.

Because this is an above-grade-level test, after the test, you'll receive information that shows how your child's score compares to that of students in the higher grades for whom the test questions were originally designed.

The SCAT is used by the Center for Talented Youth (CTY) as an above-grade-level entrance exam for students in grades 2–5. Students in grades 6-10 take the Advanced SCAT. There are 50 questions per section, 5 of which are experimental. As of publishing of this workbook the percentile ranks for the SCAT have not been updated since 1979. So, when your child takes this test, your child is being compared to a national sample of children who took the test in 1979.

Scoring is based on a three-step process in which a child's raw score is scaled based on the test version and then compared to the results of the test scores of normal students in the higher-level grade. Please keep in mind that the group of normal students took this test in 1979. So, your child's percentile ranks could be different if compared to a more recent group of test takers. The minimum scores required for qualification for the 2nd to 10th grade CTY summer courses are below:

- Grade 2 $\geq$ 430 SCAT Verbal or 435 SCAT Quantitative
- Grade 3 $\geq$ 435 SCAT Verbal or 440 SCAT Quantitative
- Grade 4 $\geq$ 440 SCAT Verbal or 450 SCAT Quantitative
- Grade 5 $\geq$ 445 SCAT Verbal or 465 SCAT Quantitative
- Grade 6 $\geq$ 450 SCAT Verbal or 470 SCAT Quantitative
- Grade 7 $\geq$ 455 SCAT Verbal or 475 SCAT Quantitative
- Grade 8 $\geq$ 460 SCAT Verbal or 480 SCAT Quantitative
- Grade 9 $\geq$ 465 SCAT Verbal or 485 SCAT Quantitative
- Grade 10 $\geq$ 470 SCAT Verbal or 490 SCAT Quantitative

SCAT Scaled Scores range from 401 to 514 depending on the level the student takes. Here are the ranges:

Elementary Level
Verbal Range = 401-471
Quantitative Range = 412-475

Intermediate Level
Verbal Range = 405-482
Quantitative Range = 419-506

Advanced Level
Verbal Range = 410-494
Quantitative Range = 424-514

This scaled score is based on the number of questions the student answers correctly out of the 50 scored questions in each section.

SCAT percentiles are used to compare students to the older population to whom the student will be compared. For example, Grade 2 students are compared to a general population of 4th graders and so on, as detailed below.

Grade 2 is compared to Grade 4
Grade 3 to Grade 5
Grade 4 to Grade 6
Grade 5 to Grade 8
Grade 6 to Grade 9
Grade 7 to Grade 12
Grade 8 to Grade 12

*Source: John Hopkins, CTY

This workbook is designed to give any child the opportunity to get familiar with SCAT question format. For a hands on approach of testing at home with mom and dad – try practicing with our workbooks. Your child will feel more confident on test day, and you will feel assured that you provided the best educational resource for your child to get ahead!

Test Taking Tips

Standardized tests are very similar in format. Most of the tests are written in multiple choice format and require a child to select the most accurate answer. Teach your child to:

1. Read the directions carefully.
2. Review all the answer choices before selecting an answer.
3. Eliminate wrong answers.
4. Trust your intuition to take an educated guess.
5. By all means, never leave a question unanswered.
6. If time permits go over your work. Double check your answers. Use scrap paper to note the questions that require extra time.

Gifted and talented tests are designed to be difficult and long. They contain advanced material and not all students will complete the entire test on time. Teach your child to budget his or her time. Go through easier questions quicker and spend more time on the challenging questions. Take a guess on those challenging questions, and come back to them if time allows. In our tutoring experience, we have had parents tell us that children underperformed in sections, in which they skipped or left questions blank. Teach your child to:

1. Do not spend too much time on a single question.
2. Do not day dream and waste time.
3. When the instructor or computer indicates that the time is almost up (i. e. less than 1 minute left) – now is the time is to guess on all the remaining questions.

As a parent, help your child pass the test. No studying or reviewing the night before the test. Make sure your child gets plenty of sleep, has a nutritious breakfast and packs a brain-stimulating snack. Most importantly keep your child calm and relaxed. Some parents are so intense and demanding, that children can even fail the test intentionally.

If you don't succeed on the first attempt try and try again. Practice makes perfect. Ask for a retake and try to improve your score the second time around. You are already on the path to success by being involved in your child's education and life.

Study Guide for the SCAT®

School and College Ability Test®

Elementary Series

SAMPLE QUESTIONS FOR GIFTED 2ND AND 3RD GRADERS

Verbal Section

60 questions

60 minutes

Directions:

Each analogy question begins with two words. These two words are related in a certain way. Below them, there are four other pairs of words lettered A, B, C, and D.

Select the pair of words that go together in the same way as the first pair of words.

1. fire : hot ::

   A. snow : cold
   B. coat : hat
   C. pants : belt
   D. eye : glasses

2. glasses : see ::

   A. glove : hand
   B. cat : pet
   C. cane : walk
   D. juice : milk

3. glove : hand ::

   A. man : uncle
   B. sock : foot
   C. buckle : belt
   D. car : truck

4. lettuce : green ::

   A. finger : toe
   B. vegetable : fruit
   C. cold : hot
   D. strawberry : red

5. reward : good ::

   A. punishment : bad
   B. say : ask
   C. green : color
   D. hammer : tool

6. happy : joyful ::

   A. silly : smart
   B. sad : depressed
   C. happy : smile
   D. emotion : face

7. car : road ::

   A. truck : bus
   B. coat : jacket
   C. boat : water
   D. sail : wind

8. moon : night ::

   A. sun : day
   B. planet : telescope
   C. summer : flowers
   D. eye : ear

9.   blue : sky ::

   A. green :grass
   B. pink : color
   C. cat : pet
   D. cane : walk

10.   straw : drink ::

   A. food : kitchen
   B. spoon : eat
   C. fork : knife
   D. ear : music

11.   morning : breakfast ::

   A. juice : milk
   B. coat : hat
   C. evening : dinner
   D. lunch : taco

12.   brother : boy ::

   A. family : happy
   B. sister : girl
   C. people : planet
   D. glasses : face

13.   apple : fruit ::

   A. carrot : vegetable
   B. cloud : rain
   C. cold : hot
   D. strawberry : red

14.   bird : feather ::

   A. finger : hand
   A. coat : hanger
   B. fish : scale
   C. man : face

15.   large : small ::

   A. truck : bus
   B. coat : thread
   C. boat : water
   D. full : empty

16.   all : none ::

   A. many : few
   B. one : two
   C. letter : number
   D. all : many

17.   open : closed ::

   A. awake : asleep
   B. silly : smart
   C. sad : depressed
   D. happy : smile

18.   bright : dark ::

   A. open : lid
   B. warm : hot
   C. rich : poor
   D. sky : sun

19. swim : summer ::

   A. glove : hand
   B. ski : winter
   C. ice : skates
   D. shoes : boots

20. fish : school ::

   A. lion : pride
   B. tiger : mammal
   C. spider : web
   D. nest : bird

21. football : field ::

   A. racquet : ball
   B. ball : net
   C. basketball : court
   D. player : referee

22. teller : bank ::

   A. building : house
   B. profession : judge
   C. ticket : money
   D. waitress : restaurant

23. dollar : bill ::

   A. penny : coin
   B. money : bank
   C. card : plastic
   D. nickel : five

24. magic : tricks ::

   A. hat : rabbit
   B. joke : funny
   C. comedy : jokes
   D. funny : surprised

25. nurse : patient ::

   A. doctor : teeth
   B. teacher : student
   C. dentist : judge
   D. person : face

26. poet : poem ::

   A. artist : painting
   B. teacher : student
   C. dentist : judge
   D. sculpture : clay

27. scissors : cut ::

   A. food : kitchen
   B. spoon : eat
   C. fork : knife
   D. ruler : yardstick

28. beagle : dog ::

   A. hamster : fish
   B. canary : bird
   C. nest : bird
   D. smile : face

29. wheel : circle ::

    A. book : rectangle
    B. shape : triangle
    C. triangle : three
    D. math : fun

30. sleepy : yawn ::

    A. ruler : measure
    B. cap : hat
    C. itchy : scratch
    D. dentist : judge

31. pound : kilogram ::

    A. quart : liter
    B. ruler : yardstick
    C. math : algebra
    D. temperature : degree

32. shoe : foot ::

    A. glove : hand
    B. coat : hat
    C. buckle : belt
    D. eye : face

33. March : spring ::

    A. December : winter
    B. Christmas : Easter
    C. ski : swim
    D. school : vacation

34. wealth : poverty ::

    A. work : money
    B. doctor : patient
    C. weather : rain
    D. sickness : health

35. fat : obese ::

    A. slender : thin
    B. tall : short
    C. hot : cold
    D. sad : happy

36. teacher : school ::

    A. school : playground
    B. coat : hanger
    C. judge : court room
    D. toy : doll

37. squirrel : acorn ::

    A. cat : pet
    B. rabbit : carrot
    C. fish : dog
    D. cabbage : seed

38. cat : kitten ::

    A. dog : puppy
    B. sheep : wool
    C. cow : milk
    D. farm : pumpkin

39. dishonest : honest ::

   A. always : never
   B. smart : genius
   C. sly : sneaky
   D. soft : fluffy

40. umpire : baseball ::

   A. glove : hand
   B. tennis : court
   C. referee : football
   D. ball : racquet

41. snake : reptile ::

   A. rabbit : carrot
   B. fish : dog
   C. frog : amphibian
   D. mammal : spider

42. sugar : sweet ::

   A. tears : salty
   B. kitchen : cook
   C. bedroom : sleep
   D. food : juice

43. month : day ::

   A. hour : minute
   B. time : clock
   C. purse : wallet
   D. count : read

44. smell : nose ::

   A. word : letter
   B. up : down
   C. sight : eyes
   D. top : side

45. go : green ::

   A. stop : red
   B. car : road
   C. train : tracks
   D. traffic : yellow

46. three : triangle ::

   A. circle : shape
   B. cone : cylinder
   C. five : odd
   D. four : square

47. hockey : puck ::

   A. stick : penalty
   B. court : ice rink
   C. baseball : ball
   D. play : rest

48. goose : flock ::

   A. fish : school
   B. frog : amphibian
   C. squirrel : acorn
   D. cow : bull

49. zebra : stripes ::

    A. cow : calf
    B. giraffe : spots
    C. zoo : exhibits
    D. lion : tiger

50. land : dirt ::

    A. ocean : water
    B. desert : hot
    C. tornado : wind
    D. toe : foot

51. book : page ::

    A. sentence : word
    B. read : listen
    C. poem : movie
    D. writer : pen

52. wheel : bike ::

    A. glove : hand
    B. tire : car
    C. buckle : seatbelt
    D. drive : truck

53. grape : purple ::

    A. rabbit : carrot
    B. rose : red
    C. garden : flowers
    D. spring : bloom

54. small : large ::

    A. enormous : huge
    B. day : month
    C. sunny : warm
    D. cold : freezing

55. fish : gills ::

    A. turtle : reptile
    B. squirrel : lungs
    C. dolphin : mammal
    D. fish : school

56. duet : two ::

    A. choir : music
    B. quartet : four
    C. violin : strings
    D. clarinet : trumpet

57. furniture : desk ::

    A. potato : tomato
    B. cat : fish
    C. tool : hammer
    D. chicken : egg

58. hide : conceal ::

    A. magic : trick
    B. show : reveal
    C. sleepy : energized
    D. top : bottom

59. necklace : neck ::

    A. glove : mitten
    B. coat : hat
    C. buckle : belt
    D. ring : finger

60. shark : swim ::

    A. dolphin : mammal
    B. parrot : feathers
    C. bird : fly
    D. cat : kitten

Quantitative Section

60 questions

60 minutes

Directions:

Each of the questions below has two parts. One part is in Column A, another part is in Column B. You must decide if one part is greater than the other or if the parts are equal. Then, select one of the three answers below:

> A if the part in Column A is greater
>
> B if the part in Column B is greater
>
> C if the two parts are equal

Column A	Column B
1. Number of tens in 59	Number of ones in 328
2. 100 + 11	12 + 99
3. one half hour	30 minutes
4. 1	0 X 5
5. The next odd number after 3	The next odd number after 5

A if the part in Column A is greater

B if the part in Column B is greater

C if the two parts are equal

	Column A	Column B
6.	700 minus 400	300 plus 400
7.	44 + 11	33 + 22
8.	Ten minus five	6
9.	$\dfrac{6}{8}$	$\dfrac{1}{2}$
10.	Number of dots below	Number of dots below

A if the part in Column A is greater

B if the part in Column B is greater

C if the two parts are equal

	Column A	Column B
11.	7,777	11,1111
12.	100 - 20	60 + 40
13.	Two times five	10
14.	$\dfrac{1}{2}$	$\dfrac{1}{3}$
15.	The distance around the triangle if each side has length 5	The distance around the rectangle if each side has length 5

| A if the part in Column A is greater |
| B if the part in Column B is greater |
| C if the two parts are equal |

	Column A	Column B
16.	49 + 1	50 - 1
17.	10 + 40 + 11	20 + 10 + 5
18.	12,657	98,231
19.	Round 13,989 to the nearest thousand	Round 13,389 to the nearest thousand
20.	Number of dots below	2 x 2

A if the part in Column A is greater

B if the part in Column B is greater

C if the two parts are equal

	Column A	Column B
21.	45,331	45,301
22.	100 + 20 + 5	100 + 20 + 7
23.	Number of days in a week	6
24.	$\dfrac{5}{3}$	$\dfrac{1}{2}$
25.	The area of the square if each side has length 10	The distance around the square if each side has length 10

A if the part in Column A is greater

B if the part in Column B is greater

C if the two parts are equal

	Column A	Column B
26.	20 x 0	20 – 10 - 10
27.	5 + 3 + 2	6 + 2 + 3
28.	909,121	990,121
29.	Round 199 to the nearest hundred	Round 206 to the nearest hundred
30.	Area of shaded region	Area of white region

A if the part in Column A is greater

B if the part in Column B is greater

C if the two parts are equal

	Column A	Column B
31.	6 quarters	1 dollar
32.	$10 - 3 - 4$	$3 + 2 + 1$
33.	5,708	5,780
34.	Next number in the sequence 10, 15, 20, 25, ...	Next number in the sequence 4,8,12,16,, ...
35.	The distance around the triangle if each side has length 10	The distance around the rectangle if each side has length 2

A if the part in Column A is greater

B if the part in Column B is greater

C if the two parts are equal

	Column A	Column B
36.	1,002	2,001
37.	13 – 7 - 2	10 – 3 - 4
38.	Three hundred one	310
39.	$\dfrac{1}{8}$	$\dfrac{1}{4}$
40.	Area of shaded region	Area of white region

A if the part in Column A is greater

B if the part in Column B is greater

C if the two parts are equal

	Column A	Column B
41.	7,398	7,308
42.	6 + 7	6 x 2
43.	Two less than 101	98
44.	$\dfrac{10}{5}$	$\dfrac{1}{2}$
45.	The distance around the triangle if each side has length 5	The distance around the rectangle if each side has length 5

A if the part in Column A is greater

B if the part in Column B is greater

C if the two parts are equal

	Column A	Column B
46.	90 - 10 + 10	90
47.	2 + 2 + 2 + 2	2 x 5
48.	19,123	91,121
49.	$\frac{8}{2}$	$\frac{2}{8}$
50.	Next number in the sequence 16, 14, 12, 10, ...	Next number in the sequence 2, 6, 10, 14, , ...

A if the part in Column A is greater

B if the part in Column B is greater

C if the two parts are equal

	Column A	Column B
51.	Number of days in a year	Number of days in a leap year
52.	11 + 1 + 2	6 + 6 + 2
53.	15,718	15,780
54.	Number of quarters in 2 dollars	6
55.	Next number in the sequence 3, 9, 12, 15, ...	Next number in the sequence 8, 12, 16, 20, ...

> A if the part in Column A is greater
> B if the part in Column B is greater
> C if the two parts are equal

	Column A	Column B
56.	3,002	3,201
57.	77+ 33	59 + 41
58.	Two more than 999	1,000
59.	Number of minutes in an hour	40
60.	$\dfrac{1}{13}$	$\dfrac{1}{2}$

Answers – Verbal Section

1. A
2. C
3. B
4. D
5. A
6. B
7. C
8. A
9. A
10. B
11. C
12. B
13. A
14. B
15. D
16. A
17. A
18. C
19. B
20. A
21. C
22. D
23. A
24. C

Answers – Verbal Section

25. B
26. A
27. B
28. B
29. A
30. C
31. A
32. A
33. A
34. D
35. A
36. C
37. B
38. A
39. A
40. C
41. C
42. A
43. A
44. C
45. A
46. D
47. C
48. A

Answers – Verbal Section

49.	B
50.	A
51.	A
52.	B
53.	B
54.	D
55.	B
56.	B
57.	C
58.	B
59.	D
60.	C

Answers – Quantitative Section

1. B
2. C
3. C
4. A
5. B
6. B
7. C
8. B
9. A
10. A
11. B
12. B
13. C
14. A
15. B
16. A
17. A
18. B
19. A
20. C
21. A
22. B
23. A
24. A

Answers – Quantitative Section

25. A
26. C
27. B
28. B
29. C
30. C
31. A
32. B
33. B
34. A
35. A
36. B
37. A
38. B
39. B
40. A
41. A
42. A
43. A
44. A
45. C
46. C
47. B
48. B

Answers – Quantitative Section

49. A
50. B
51. B
52. C
53. B
54. A
55. B
56. B
57. A
58. A
59. A
60. B

My Notes:

My strengths:

- ✓
- ✓
- ✓
- ✓

Topics to work on:

- ✓
- ✓
- ✓
- ✓

www.ingramcontent.com/pod-product-compliance
Lightning Source LLC
Chambersburg PA
CBHW081421220526
45467CB00009B/2771